T0225193

Nonlinear Control Systems Using MATLAB®

Nonlinear Control Systems Using MATLAB®

Authored by
Mourad Boufadene

CRC Press
Taylor & Francis Group
Boca Raton London New York

CRC Press is an imprint of the
Taylor & Francis Group, an **Informa** business

CRC Press
Taylor & Francis Group
6000 Broken Sound Parkway NW, Suite 300
Boca Raton, FL 33487-2742

First issued in paperback 2021

© 2019 by Taylor & Francis Group, LLC
CRC Press is an imprint of Taylor & Francis Group, an Informa business

No claim to original U.S. Government works

Printed on acid-free paper
Version Date: 20180816

ISBN-13: 978-1-138-35955-0 (hbk)
ISBN-13: 978-1-03-209473-1 (pbk)

**Visit the Taylor & Francis Web site at
http://www.taylorandfrancis.com**

**and the CRC Press Web site at
http://www.crcpress.com**

Contents

Preface

This book introduces nonlinear control systems for control engineering and science to graduate, undergraduate students and researchers; it targets control engineering students who do not like to do not have time to derive and prove mathematical results for nonlinear control systems. It can be serve as a text book for nonlinear control systems, especially for feedback linearization techniques which is a common approach in controlling nonlinear systems.

The development of computer software for nonlinear control systems has provided many benefits for teaching, research, and the development of control systems design. MATLAB® is considered the dominant software platforms for linear and nonlinear control systems analysis. This book contains a MATLAB-based program that helps teachers, students, and researchers for finding feedback linearization controller for a specific nonlinear class of systems.

Chapter 1 is an introduction to theoretical aspects of nonlinear feedback linearization techniques. We use simple and complex examples to better illustrate the method on how to find the feedback linearization controller for single and multi input output systems. In section 1.9, a MATLAB-based program is developed and used to find feedback linearization control using symbolic MATLAB libraries for a special class of nonlinear systems.

Chapter 2, the concept of structure variable control will be introduced in intuitive way, with illustrative examples that makes the reader familiar with the theory of sliding mode control in an easy and simple way. After having defined the control law, the parameters of the controller should be identified, hence a simple method is presented to find those parameters. In section 2.4 a MATLAB-based function is developed to find the sliding mode surfaces and controllers using symbolic MATLAB library for SISO and MIMO nonlinear dynamical systems that could be written in a special form.

This textbook provides an easy way to learn nonlinear control systems such as feedback linearization technique and sliding mode control (structrue variable control) which are the most used techniques in nonlinear control dynamical systems; therefore, teachers, students and researchers are all in need of learning how to handle such techniques since they are too difficult for them to handle such nonlinear controllers, especially for more complicated systems such as induction motor, satellite, and vehicles dynamical models. Thus, this document is an excellent resource for learning the principle of feedback linearization and sliding mode techniques in an easy and simple way. This book:

(a) provides a briefs description of the feedback linearization and sliding mode control strategies

(b) includes a simple method on how to determine the right and appropriate controller (P-PI-PID) for feedback linearization control strategy

(c) provides a MATLAB symbolic based programs that can solve the Lie derivative of any nonlinear dynamical system that could be written in a special form

(d) provides a symbolic MATLAB-based function for finding the feedback linearization, and sliding mode controllers are developed and tested using several examples

(e) introduces a simple method for finding the approximate sliding mode controller parameters

The relative degree of the system which can be found using the provided MATLAB code can be used to determine the degree of the Sliding Mode Surface, which is a very important scalar for those who are working with Variable Structure Control.

There are many examples used in the last chapter with a modified MATLAB program such as:

(a) A nonlinear dynamical model of a pendulum system

(b) A nonlinear dynamical model of Ven der pol system

(c) A nonlinear dynamical mode of an aircraft

(d) A nonlinear dynamical model of an induction motor

(e) A nonlinear dynamical model of a Permanent magnet synchronous motor.

Where the program used to construct the nonlinear controller uses symbolic computations such that the user should provide the program with the necessary functions $f(x)$, $g(x)$ and $h(x)$ using the symbolic library.

Note: Make sure the function syms exists on your MATLAB by typing help syms on MATLAB.

MATLAB® is a trademark of The MathWorks, Inc. and is used with permission. The MathWorks does not warrant the accuracy of the text or exercises in this book. This book's use or discussion of MATLAB® software or related products does not constitute endorsement or sponsorship by The MathWorks of a particular pedagogical approach or particular use of the MATLAB® software.

--

MATLAB® is a registered trademark of The MathWorks, Inc. For product information, please contact:

The MathWorks, Inc.
3 Apple Hill Drive
Natick, MA, 01760-2098 USA
Tel: 508-647-7000
Fax: 508-647-7001
E-mail: info@mathworks.com
Web: www.mathworks.com

CHAPTER 1

Feedback Linearization Control

A FEEDBACK LINEARIZATION is a common approach used in controlling nonlinear systems. The approach involves coming up with a transformation to the nonlinear system into equal linear system that could be controlled easily using a new input control. Feedback linearization could be applied to nonlinear systems of the form:

$$\dot{x} = f(x) + g(x)u$$
$$y = h(x) \tag{1.1}$$

Where $x \in \Re^n$ is the state vector, $x \in \Re^p$ is the vector of inputs, and $y \in \Re^m$ is the vector of outputs. The goal is to develop a control input:

$$u = D(x)^{-1}\left[-A(x) + v\right] \tag{1.2}$$

That renders a linear input-output map between the new input v and the output. An outer-loop control strategy for the resulting linear control system can then be applied.

1.1   FEEDBACK LINEARIZATION OF SISO SYSTEMS

Let's consider the following SISO nonlinear dynamical system described by:

$$\dot{x} = f(x) + gu$$
$$y = h(x) \tag{1.3}$$

Where $x = [x_1, x_2, ..., x_n]$ is the state vecotr; u input vector to the system; $f(x)$ and $g(x)$ are infinite differentiable vectors; hence the input-output feedback control law u is defined for SISO nonlinear systems in (1.6) by the following relation:

$$u = \frac{1}{L_g L_f^{r-1} h(x)} (-L_f^r h(x) + v) \qquad (1.4)$$

Where r is the relative degree of the system; $L_g L^{r-1} f(x)$ and $L_f^r h(x)$ are the lie derivative of $h(x)$ along the vector fields $g(x)$ and $f(x)$, respectively. The notion of the lie derivatives and how its calculated will be given in the subsequent sections. Thus, the overall system control law u and nonlinear system (1.6) combined in series yields to a linear system with order r (relative degree) from the new input v to output y

$$y^r = v \qquad (1.5)$$

Where v is the new input vector to the system; that will be used as new controlled input.

RELATIVE DEGREE OF SISO SYSTEMS

1. $r = n$ the system admits an exact state linearization

2. $r \leq n$ the system admits partial feedback linearization

3. $r > n$ the system does not admit an input output feedback linearization

1.2 FEEDBACK LINEARIZATION OF MIMO SYSTEMS

Feedback linearization control for MIMO system is applied to dynamical systems that have the following form:

$$\dot{x} = f(x) + \sum_{i=1}^{m} g_i(x) u_i$$

$$y_1 = h_1(x)$$
$$y_2 = h_2(x) \qquad (1.6)$$

$$\vdots$$

$$y_n = h_n(x) \qquad (1.7)$$

Equation 1.6 is called a square system, where the number of inputs is equal to the number of outputs.

The control law for MIMO systems can be obtained using lie derivative.

$$\begin{bmatrix} y_1^{r_1} \\ y_2^{r_2} \\ ... \\ y_m^{r_m} \end{bmatrix} = A(x) + D(x)u \tag{1.8}$$

Where

$$A(x) = \begin{bmatrix} L_f^{r_i} h_j(x) \\ .... \\ L_f^{r_m} h_m(x) \end{bmatrix} \tag{1.9}$$

$$D(x) = \begin{bmatrix} L_{g_1} L_f^{r_1-1} h_1(x) & . & L_{g_m} L_f^{r_1-1} h_1(x) \\ & .... & \\ L_{g_1} L_f^{r_1-1} h_m(x) & . & L_{g_m} L_f^{r_m-1} h_m(x) \end{bmatrix} \tag{1.10}$$

Where $D(x)$ is called the decoupling matrix of the system, and it is not always nonsingular, and therefore the linearization control is then achieved if and only if the decoupling matrix is invertible

$L_f^{r_i} h_j(x)$ The lie derivative of $h(x)$ along the vector field $f(x)$

$L_{g_m} L_f^{r_m-1} h_m(x)$ The lie derivative of $h(x)$ along the vector field $g(x)$

Therefore, the linearization control law is then written in compact form as:

$$u = D(x)^{-1} [-A(x) + v] \tag{1.11}$$

Where v is the new input control vector which has the same dimension as u.

RELATIVE DEGREE OF MIMO SYSTEMS

1. $r = r_1 + r_2 + ...r_m = n$ then the system admits an exact

2. $r = r_1 + r_2 + ...r_m \leq n$ then the system admits partial feedback linearization

3. $r > n$ the system does not admit an input output feedback linearization for both types of systems

1.3 RELATIVE DEGREE

The relative degree of a dynamic systems is obtained by differentiating the output y until the input u appears

Theorem 1 *A nonlinear system is said to have a relative degree r at x iff:*

$$L_g L_f^{r-1} h(x) \neq 0 \quad for \quad r \neq n \quad and \quad r \leq n$$

Where $L_f h$ is the lie derivative of h along the vector field $f(x)$

$$L_f h(x) = \sum_{n}^{i=1} \frac{\partial h(x)}{\partial x_i} f_i(x)$$

Where $L_f h$ is the derivative of h along the vector field $g(x)$

$$L_g h(x) = \sum_{n}^{i=1} \frac{\partial h(x)}{\partial x_i} g_i(x)$$

1.4 MATLAB® PROGRAM DESCRIPTION

The technique of an input-output feedback linearization is used for many control applications, which transforms a complicated and coupled nonlinear system into an equivalent linear system that could be controlled using linear control system tools, such as pole placement technique. Thus, to find such control law a MATLAB script is developed to facilitate the tedious computations and avoiding errors due to a complexity of the system.

1.5 PROBLEM FORMULATION

Let's consider the following nonlinear mathematical system representation:

$$\begin{cases} \dot{x} = & f(x) + \sum_m^{i=1} g_i(x) u_i \\ y_j = & h_j(x) \end{cases} \tag{1.12}$$

Where $x \in \Re^n$ is the state vector; $x \in \Re^n$ is the input vector of the system; $h_j(x)x \in \Re^j$ Is an analytical function of x; $f(x)$ and $g(x)$ are infinitely differentiable vector fields.

1.5.1 Programme flow chart

The flow chart of Figure 1 gives a detailed explanation of how the program works; therefore, the following steps are used to obtain the nonlinear control law using the developed program:

(a) Introduce the nonlinear functions $f(x), g(x)$ and $h(x)$ as a symbolic functions to the program

(b) Find the lie derivative along the vector fields $f(x)$ and $g(x)$

(c) If the lie derivative along the vector fields $g(x) = 0$ and the number of differentiation is less than n (the system state number), then set $f(x) = L_f h(x)$ and repeat step 2

(d) If the lie derivative along the vector fields $L_g h(x) \neq 0$ and the number of differentiation is or equal to n (the system state number), then prints the lie derivative along the vector fields $f(x)$ and $g(x)$ as well as the differentiation number r, which is the relative degree, which could be used to identify the controller type

(e) If the lie derivative along the vector fields $L_g h(x) = 0$ and the number of differentiation is equal or greater than n (the system state number), then the system does not admit an input output feedback linearization

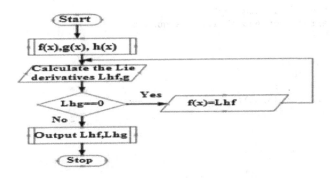

Figure 1.1 The flow chart of the proposed algorithm

1.6   MATLAB CODE OF LIE DERIVATIVE

The aim of feedback linearization is to produce a transformation system whose states are the output y and its derivatives. Therefore, the lie derivative is then used, and since it's an important tool for analysis and synthesis, a MATLAB program has been developed for any class of nonlinear system that could be written in the form of an equation (1.12). The two programs work as follows:

I The *LieDerivative*(h, x) program finds the partial derivative of $h(x)$ along x; as described by the following equation: $\dot{f}(x) = \frac{\partial h(x)}{\partial x}\dot{x}$; thus, the user should provide the program with the vector field

II The *solvelieder*(Lh, fx, g) gives the lied derivatives of the vector fields $f(x)$ and $h(x)$ along the vector fields $f(x)$ and $g(x)$ respectively. Where the user has to provide the program with the function $f(x)$ and $g(x)$ as well as the output derivative from the function *LieDerivative*(h, x)

```
function df=LieDerivative(h,x)
% The LieDerivaive MATLAB function is used
% to find the jacobian vector of a given output
% h(x)      : Is the output function
% x         : The state vector
% df        : The jacobian of h along x

if nargin<2 & nargin==0
    error('not enough input argument');
end
df=[];
n=length(x);
for ii=1:n
    xx=x(ii);
    dff=diff(H,xx);
    df=[df,dff];
end
df;
end


function [lhf lhg]=solvelieder(Lh,fx,g)
%   The solvelieder MATLAB function is used to find
%   the lie derivatives of the functions f(x) and g(x)
%   along the vector field h(x)
%   Lh    : The jacobian vector of h along x
%   fx    : The function f(x)
%   g     : The input function g(x)
```

```
LHg=[];
lhf=Lh*fx;
[n,b]=size(G);
for ii=1:b
    Lgh=Lh*G(:,ii);
    LHg=[LHg,Lgh];
end
lhg=LHg;
```

1.6.1 Example for SISO systems

Find the lie derivative of the following system with $h(x) = x_1$

$$\begin{cases} \dot{x}_1 = & x_2 \\ \dot{x}_2 = & -sin(x_1) + u \end{cases} \tag{1.13}$$

Hand Calculation The lie derivative of any nonlinear dynamical systems that has the form of Equation 1.1 is obtained by differentiating the output of the system; hence, the first lie derivative of Equation 1.13 is given by:

$$\dot{y} = \frac{\partial h(x)}{\partial x}\dot{x}$$
$$= \begin{bmatrix} 1 & 0 \end{bmatrix} (f(x) + gu) \tag{1.14}$$

Therefore, from Equation 1.14 the first lie derivative is given by:

$$\dot{y} = \begin{bmatrix} 1 & 0 \end{bmatrix} \begin{pmatrix} x_2 \\ -sin(x_1) + u \end{pmatrix} \tag{1.15}$$

$$\dot{y} = L_f^1 h(x) = x_2 \tag{1.16}$$

MATLAB Computation The first lie derivative of equation (1.13) is then obtained using MATLAB program described on the previous section as follows:

```
syms x1 x2 u  % symbolic presentation
fx=[x2;-sin(x1)];
g=[0;1];
x=[x1 x2];H=x1;
LH=Lie_Derivative(H,x)
[lhf lhg]=solve_lie_der(LH,fx,g)
```

Program output

```
LH = [ 1, 0]
lhf = x2
lhg =0
```

The first lie derivative of the dynamical system in Equation 1.13 is displayed using MATLAB as $L_f h(x) = lhf$ and $L_g h = lhg$.

1.7 CONTROLLER TYPES

After having defined the matrix $A(x)$ and $D(x)$ as well as the relative degree of the system, which has a significant importance of finding the appropriate type of controller, and since that the transfer function of the linearized system is given in terms of the relative degree are as:

$$\frac{y}{u} = \frac{1}{s^r} \qquad (1.17)$$

Hence, the controller type will be defined in an appropriate manner using values of the relative degree are as:

- If r=1 then the appropriate controller is a Proportional P

- If r=2 then the appropriate controller is Proportional Derivative PD

- If r=3 or greater than that, the appropriate controller is Proportional-Integra-Derivative PID

1.8 FEEDBACK LINEARIZATION CONTROLLER EXAMPLES

An illustrative examples will be used to show users how to use MATLAB code for finding the nonlinear control law. Therefore, for better readability of the program, nonlinear dynamical systems are used.

1.8.1 Example for SISO systems

The nonlinear dynamical model of a pendulum is written in compact form as

$$\dot{x} = f(x) + gu = \begin{pmatrix} x_2 \\ -sin(x_1) \end{pmatrix} + \begin{bmatrix} 0 \\ 1 \end{bmatrix} u \qquad (1.18)$$

Find the control law u that linearized model with $h(x) = x_2$?

1.8.1.1 Solution Using Hand

Following the steps mentioned in the flow chart of Figure 1, then the control law will be obtained by differentiating the output h till u appears as:

$$\dot{y}(x) = \frac{\partial h(x)}{\partial x}\dot{x} = \frac{\partial h(x)}{\partial x}(f(x) + gu) \quad (1.19)$$

Where

$$L_f^1 h(x) = \frac{\partial h(x)}{\partial x}f(x) = [0 \ 1]\begin{bmatrix} x_2 \\ -sin(x_1) \end{bmatrix} = -sin(x_1) \quad (1.20)$$

$$L_g^1 h(x) = \frac{\partial h(x)}{\partial x}g(x) = [0 \ 1]\begin{bmatrix} 0 \\ 1 \end{bmatrix} = 1 \quad (1.21)$$

Where the superscript on Equation 1.17 denotes the relative degree of the system which is 1, and $L_f h$, $L_g h$ are the lie derivatives of the vector field h along the vector fields f(x) and g(x), respectively. Furthermore, the control law u is obtained using Equation 1.9 as:

$$u = sin(x_1) + v \quad (1.22)$$

1.8.1.2 Solution Using MATLAB

Equation 1.22 is then obtained using MATLAB script presented at the end of this document by:

```
The Nonlinear systems should be written in the following
form——Feedback Linearization Controller for a class of
          Nonlinear systems——State space equations
          x=f(x)+g(x)u
```

```
Parmeters

parameters =

   Empty string: 1—by—0


Number of states:=2
Number of inputs:=1
The vector f(x):=[x2;—sin(x1)]
The vector g(x):=[0;1]
The output variables:=x2
The relative degree of  h1
equal:=1
```

The feedbacklinearization controller Uc:

Uc =

u1 + sin(x1)

By comparing the control law obtained in Equation 1.22 using hand calculation with that obtained using MATLAB script code shows the effectiveness of the program for single input-single output system.

1.8.2 Example for MIMO systems

In order to show the effectiveness of the program for a system that has a multi-input-multi-output, the following nonlinear dynamcial system will be used:

$$\dot{x} = f(x) + gu = \begin{bmatrix} x_1 + x_1 x_2 \\ -sin(x_1) \end{bmatrix} + \begin{bmatrix} 1 & 0 \\ 0 & 1 \end{bmatrix} \begin{bmatrix} u_1 \\ u_2 \end{bmatrix} \quad (1.23)$$

The output vector of the system is given by: $h(x) = \begin{bmatrix} h_1 \\ h_2 \end{bmatrix} = \begin{bmatrix} x_1 \\ x_2 \end{bmatrix}$

1.8.2.1 Soluting Using Hand

Following the steps mentioned in the flow chart of Figure 1; then the control law will be obtained by differentiating the output h till u appears as:

$$\dot{y}(x) = \frac{\partial h(x)}{\partial x} \dot{x} = \frac{\partial h(x)}{\partial x} (f(x) + gu) \quad (1.24)$$

Where

$$L_f^1 h_1(x) = \frac{\partial h(x)}{\partial x} f(x) = [1 \ 0] \begin{bmatrix} x_1 + x_1 x_2 \\ -sin(x_1) \end{bmatrix} = x_1 + x_1 x_2 \quad (1.25)$$

$$L_f^1 h_2(x) = \frac{\partial h(x)}{\partial x} f(x) = [0 \ 1] \begin{bmatrix} x_1 + x_1 x_2 \\ -sin(x_1) \end{bmatrix} = -sin(x_2) \quad (1.26)$$

$$L_g^1 h_1(x) = \frac{\partial h(x)}{\partial x} f(x) = [1 \ 0] \begin{bmatrix} 1 & 0 \\ 0 & 1 \end{bmatrix} = 1 \quad (1.27)$$

$$L_g^1 h_2(x) = \frac{\partial h(x)}{\partial x} f(x) = [0 \ 1] \begin{bmatrix} 1 & 0 \\ 0 & 1 \end{bmatrix} = 1 \quad (1.28)$$

Where the superscript on Equation 1.10 denotes the relative degree of the system which is 1, and $L_f h$, $L_g h$ are the lie derivatives of the vector field h along the vector fields f(x) and g(x), respectively. Furthermore, the control law u is obtained using Equation 1.11 as:

$$\begin{bmatrix} u_1 \\ u_2 \end{bmatrix} = \begin{bmatrix} 1 & 0 \\ 0 & 1 \end{bmatrix} \begin{bmatrix} -L_f^1 h_1(x) + v_1 \\ -L_f^1 h_2(x) + v_2 \end{bmatrix} \tag{1.29}$$

1.8.2.2 Solution Using MATLAB

Equation 1.29 is then obtained using the MATLAB script presented at the end of that document as shown:

The Nonlinear systems should be written in the following form——Feedback Linearization Controller for a class of
          Nonlinear systems——State space equations
          x=f(x)+g(x)u

```
Parmeters

parameters =

   Empty string: 1—by—0

Number of states:=2
Number of inputs:=2
The vector f(x):=[x1+x1*x2;—sin(x1)]
The vector g(x):=[1 0;0 1];
The output variables:=[x1;x2]
The feedback linearization controller Uc:
inv(Lhg)*(—Lhf+u)
The relative degree of  h1
equal:=1
The relative degree of  h2
equal:=1

Lhf =

x1 + x1*x2
  —sin(x1)


Lhg =

[ 1,  0]
[ 0,  1]
```

By comparing the control law obtained in Equation 1.29 using hand calculation with that obtained using MATLAB script code shows the effectiveness of the program for multi-input-multi-output systems.

1.9 MATLAB BASED FUNCTION: FEEDBACK LINEARIZATION

FEEDBACK LINEARIZATION problems are then solved using MATLAB-based function; and for better readability of the programmes illustrated on the previous chapters, we will write them on another manner that simplify the use for users; hence, the program that should be used by the user will be written as:

1.10 MATLAB TEST FUNCTION

```
clear all;clc
disp('────────────────────────────');
disp('The Nonlinear systems should    ');
disp('be written in the following form    ');
disp('State space equations    x=f(x)+g(x)u');
disp('────────────────────────────');
% The your system contains
% Input the extra parameters
par=input('Parameters  ','s');
eval(sprintf('syms %s',par));
parameters=sprintf('%s',par)
%% Declare how many states and inputs
n=input('Number of states:=');
nin=input('Number of inputs:=');
x=sym(zeros(1,n));
u=sym(zeros(1,nin));
for j=1:n
    eval(sprintf('syms x%d',j))
    x(:,j)=sprintf('x%d',j);
end
for k=1:nin
    eval(sprintf('syms u%d',k));
    u(:,k)=sprintf('u%d',k);
end
% Enter the functions from the keyboard
f=input('The vector f(x):=','s');
g=input('The vector g(x):=','s');
Hc=input('The output variables:=','s');
%Represent all the functions
%f(x), g(x) and h(x) on a symbolic format
fx=sym(f);
g=sym(g);
Hc=sym(Hc);  %
```

```
%   Use the inoutfeedbacklinearization.m
%   programm to generate the desired functions
[Lhf Lhg]=inoutfeedbacklinearization(fx,g,Hc,x)
```

"newlabel1.101537

1.10.1   Feedback linearization MATLAB function

Thus the *inoutfeedbacklinearization.m* function is then written using a nested MATLAB functions instead of using the programs separately as shown on the previous chapters; therefore the whole code is given by:

```
function [Lhf Lhg]=inoutfeedbacklinearization(fx,g,h,x)
% The function inoutfeedbacklinearization is used to find the
% the feedbacklinearization control law for SISO and MIMO
%  nonlinear systems using symbolic MATLAB library;
% The user should provide the program with the following
% inputs
% fx   : The system function f(x)
% g    : The system output function g(x)
% h(x) : The vector of outputs h(x)=[x1;x2,   ;xn]
% x    : The state vector x=[x1,x2,   ,xn]
%  After having provided the program the necessary input
%  functions
% The program will output the following variables
% Lhf  : The lie derivative of h(x) along the function f(x)
% Lhg  : The lie derivative of h(x) along the function g(x)
%         which is called the decoupling matrix
% u    : The vector of inputs u=[u1;u2;...;un]
% The control law will be given by the following formula
% u= inv(Lhg)*(—Lhg+v)

if nargin <4
error('Not enough input argument');
end
k=1;
Lhg=[];Lhf=[];
nb=length(h);
while k<length(h)+1
h1=h(k);
for i=1:nb+1
% this Lie derivative function
df=Lie_Derivative(h1,x);
% solve for the g
[lhf lhg]=solve_lie_der(df,fx,g);
[n b]=size(lhg);
for ii=1:n
  d=any(lhg(ii,:)~=0);
end
if d==1;
disp(['The relative degree of  h',num2str(k)]),
```

```
disp(['equal:=',num2str(i)]);
break;
else
    h1=lhf;
end
if i==nb+1 && d==0
disp(['The system dose not admit NFL']);
return;
end
end
Lhg=[Lhg;lhg];
Lhf=[Lhf;lhf];
k=k+1;
end
function df=LieDerivative(H,x)
% The LieDerivative MATLAB function is used
% to find the jacobian vector of a given output
% H(x)    : Is the output function
% x       : The state vector
% df      : The jacobian of h along x

if nargin<2 & nargin==0
    error('not enough input argument');
end
df=[];
n=length(x);
for ii=1:n
    xx=x(ii);
    dff=diff(H,xx);
    df=[df,dff];
end
df;
end
function [lhf lhg]=solvelieder(LH,fx,G)
%  The solvelieder MATLAB function is used to find
%  the lie derivatives of the functions f(x) and g(x)
%  along the vector field h(x)
%  LH    : The jacobian vector of h along x
%  fx    : The function f(x)
%  g     : The input function g(x)

LHg=[];
lhf=LH*fx;
[n,b]=size(G);
for ii=1:b
    Lgh=LH*G(:,ii);
    LHg=[LHg,Lgh];
end
lhg=LHg;
end
end
```

In order to use the two programmes, the user should follow the steps:

- Save the function *inoutfeedbacklinearization.m* given on the code above as shown on Figure 1.2

```
1 function [Lhf Lhg]=inoutfeedbacklinearization(fx,g,h,x)
46     % the vector field func(x)
47     % H        : the output vector
48    -% x        : the state varibales
49
50 -   if nargin<2 & nargin==0
51 -       error('not enough input argument');
52 -   end
53 -   df=[];
54 -   n=length(x);
55 -   for ii=1:n
56 -       xx=x(ii);
57 -       dff=diff(H,xx);
58 -       df=[df,dff];
59 -   end
60 -   df;
61 -   end
62     function [lhf lhg]=solvelieder(LH,fx,G)
63     % This equation solves the Lie derivatvies that is mulitiplied by
64     % The output vectors Lhg=Lh*g*u where
65     % LH       : The Lie derivatvies of h(x) along the vector field fx
66     % fx       : The vector field f(x) that describe the system
67    -% G        : The vector filed of the input
68 -   LHg=[];
69 -   lhf=LH*fx;
70 -   [n,b]=size(G);
71 -   for ii=1:b
72 -       Lgh=LH*G(:,ii);
```

Figure 1.2 inoutfeedbacklinearization MATLAB Code

- The call program where the user has to provide the program with necessary input functions is given as

```
clear all;clc
disp('————————————————————————————————');
disp('The Nonlinear systems should be written in the
        following form');
disp('—Feedback Linearization Controller for a class of
        Nonlinear systems—');
disp('          State space equations    x=f(x)+g(x)u
');
disp('————————————————————————————————');
%% Declare how many states and inputs
%  The your system contains
%  Inout the extra parameters
par=input('Parameters  ','s');
eval(sprintf('syms %s',par));
parameters=sprintf('%s',par)
n=input('Number of states:=');
```

```
nin=input('Number of inputs:=');
x=sym(zeros(1,n));
u=sym(zeros(1,nin));
for j=1:n
    eval(sprintf('syms x%d',j))
    x(:,j)=sprintf('x%d',j);
end
for k=1:nin
    eval(sprintf('syms u%d',k));
    u(:,k)=sprintf('u%d',k);
end
%% Enter the functions from the keyboard
f=input('The vector f(x):=','s');
g=input('The vector g(x):=');
Hc=input('The output variables:=','s');
%% Represent all the functions f(x), g(x) and h(x) on a
%% symbolic format
fx=sym(f);
%g=sym(g);
Hc=sym(Hc);   %
%% Use the inoutfeedbacklinearization.m programm to generate
%% the desired functions
disp(['The feedbacklinearization controller Uc:']);
disp(['inv(Lhg)*(-Lhf+u)'])
[Lhf Lhg]=in_out_feedback_linearization(fx,g,Hc,x)
```

1.11   ILLUSTRATIVE EXAMPLES

1.11.1   Aircraft altitude dynamics

Let us determine the associated internal dynamics of the aircraft by defining the state x as $[\alpha, \dot{\alpha}h, \dot{h}]$, thus, the equations of motion can be written as:

$$\dot{x}_1 = x_2$$
$$\dot{x}_2 = -4x_2 - 4x_1 + 3E$$
$$\dot{x}_3 = x_4 \tag{1.30}$$
$$\dot{x}_4 = 6x_1 - E \tag{1.31}$$

Where E is the input vector, $\alpha = x_1$ and $\alpha = x_2$ are the angle and angular speed of the aircraft, respectively; h is the altitude and $\dot{h}$ is the altitude speed see Figure 1.3. By applying the MATLAB-based

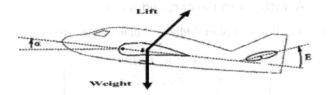

Figure 1.3 Dynamic charactercistics of an aircraft

program the feedback linearization controller of the aircraft is given below:

```
The Nonlinear systems should be written in the following
form—Feedback Linearization Controller for a class of
          Nonlinear systems—State space equations
          x=f(x)+g(x)u
```

```
parameters =

    Empty string: 1—by—0

Number of states:=4
Number of inputs:=1
The vector f(x):=[x2;-4*x2-4*x1;x4;6*x1]
The vector g(x):=[0;3;0;-1];
The output variables:=x3
The feedbacklinearization controller Uc:
inv(Lhg)*(-Lhf+u)
The relative degree of  h1
equal:=2

Lhf =

6*x1


Lhg =

-1
```

From the output of the program shown above and Equation 1.9 the new control vector v is given by:

$$E = 6x_1 - v \qquad (1.32)$$

1.11.2 Asynchronous motor speed control

The dynamical model of an induction machine is given by:

$$
f(x) =
\begin{bmatrix}
-\gamma x_1 + \frac{K}{T_r} x_3 + pK x_5 x_4 \\
-\gamma x_2 + \frac{K}{T_r} x_4 - pK x_5 x_3 \\
\frac{K}{T_r} x_1 - \frac{1}{T_r} x_3 - px_5 x_4 \\
\frac{K}{T_r} x_2 - \frac{1}{T_r} x_4 + px_5 x_3 \\
\frac{pM}{J_m L_r}(x_3 x_2 - x_4 x_1) - \frac{f_m}{J_m} x_5 - \frac{T_l}{J_m}
\end{bmatrix}
\tag{1.33}
$$

The control input matrix is defined by:

$$
g =
\begin{bmatrix}
\frac{1}{\sigma L_s} & 0 & 0 & 0 \\
0 & \frac{1}{\sigma L_s} & 0 & 0
\end{bmatrix}^T
\tag{1.34}
$$

Where the parameters σ, γ, T_r are defined as: $\sigma = 1 - \frac{M^2}{L_r L_s}$, $\gamma = \frac{R_s}{\sigma L_s} + \frac{R_r M^2}{s L_r^2}$ $T_r = \frac{L_r}{R_r}$ Where σ is the scattering coefficient, T_r is the time constant of the rotor dynamics, J_m is the rotor inertia, f_m is the mechanical viscous damping, p is the number of pole pairs, τ_d is the unknown external load torque which will be estimated as well.

The state variables $i_{sd}, i_{sq}, \phi_{rd}, u_{sd}, u_{sq}$, are the stator currents, rotor flux linkages, stator terminal voltages respectively. and for the parameters L_s, L_r, M, R_r are the rotor inductance, stator inductance, mutual inductance, stator resistance, and rotor resistance, respectively

$$
u = [u_{s\alpha} \quad u_{s\beta}] \quad x = [i_{s\alpha} \quad i_{s\beta} \quad \phi_{r\alpha} \quad \phi_{r\beta} \quad \Omega]
\tag{1.35}
$$

The output vector is then given by:

$$
h(x) =
\begin{bmatrix}
x_3^2 + x_4^2 \\
\Omega = x_5
\end{bmatrix}
\tag{1.36}
$$

Using the program, the output will be:

```
The Nonlinear systems should be written in the following
form—Feedback Linearization Controller for a class of
        Nonlinear systems—State space equations
        x=f(x)+g(x)u
```

```
Parameters  Tl gamma K Tr p fm Jm Lr M Ls sigma
```

```
parameters =

Tl gamma K Tr p fm Jm Lr M Ls sigma

Number of states:=5
Number of inputs:=2
The vector f(x):=[-gamma*x1+(K/Tr)*x3+p*K*x5*x4;
                  -gamma*x2+(K/Tr)*x4-p*K*x5*x3;
                  (M/Tr)*x1-(1/Tr)*x3-p*x5*x4;(M/Tr)*x2
                  -(1/Tr)*x4+p*x5*x3;
                  p*(M/(Jm*Lr))*(x3*x2-x4*x1)
                  -(fm/Jm)*x5-Tl/Jm]
The vector g(x):=[1/(sigma*Ls) 0;0 1/(sigma*Ls);0 0;0 0;0 0]
The output variables:=[x3^2+x4^2;x5]
The feedbacklinearization controller Uc:
inv(Lhg)*(-Lhf+u)
The relative degree of  h1
equal:=2
The relative degree of  h2
equal:=2


Lhf =

((4*x3)/Tr - (2*M*x1)/Tr)*(x3/Tr + p*x4*x5 - (M*x1)/Tr)
 - ((4*x4)/Tr - (2*M*x2)/Tr)*(p*x3*x5 - x4/Tr + (M*x2)/Tr)
 + (2*M*x3*((K*x3)/Tr - x1*gamma + K*p*x4*x5))/Tr
 - (2*M*x4*(x2*gamma - (K*x4)/Tr + K*p*x3*x5))/Tr ;

(fm*(Tl/Jm + (fm*x5)/Jm + (M*p*(x1*x4 - x2*x3))/(Jm*Lr)))/Jm
 - (M*p*x3*(x2*gamma - (K*x4)/Tr + K*p*x3*x5))/(Jm*Lr)
 - (M*p*x4*((K*x3)/Tr - x1*gamma + K*p*x4*x5))/(Jm*Lr)
 - (M*p*x1*(p*x3*x5 - x4/Tr + (M*x2)/Tr))/(Jm*Lr)
 - (M*p*x2*(x3/Tr + p*x4*x5 - (M*x1)/Tr))/(Jm*Lr)


Lhg =

[     (2*M*x3)/(Ls*Tr*sigma),     (2*M*x4)/(Ls*Tr*sigma)]
[  -(M*p*x4)/(Jm*Lr*Ls*sigma),  (M*p*x3)/(Jm*Lr*Ls*sigma)]
```

CHAPTER 2

Sliding Mode Control

I N practical control problems there will be a mismatch between the real plant and its developed mathematical model used for control purposes. These mismatches arise from uncertainties due to changes in the plant parameters or due to external disturbances.

The performance of any nonlinear dynamical systems that have such uncertainties becomes a challenging task for control engineering; a robust control method is developed to solve this problem which so-called **Sliding mode control** or **Structure variable control** [8] is used to construct the sliding mode surface; and, therefore the sliding mode controller for a class of nonlinear dynamical systems that can be written as:

2.1 SLIDING MODE CONTROL THEORY

Let's consider the following nonlinear dynamical system:.

$$\dot{x} = f(x,t) + g(x,t)u \tag{2.1}$$

$$y = h(x,t) \tag{2.2}$$

Where y and u denote the output and input variable, x is the state vector. The sliding mode control makes the output y tracks a desired input reference y_r, so that the system motion is kept on the manifold S, which is defined by:

$$S = \sigma(x,t) = 0 \tag{2.3}$$

S represents the sliding mode surface; and, therefore, the state variables has to be brought to the siding surface in order to obtain the control law u; which is given by:

u_{eq} : is the equivalent control vector, which is obtained from the derivative of Equation 2.3

u_n : is the discontinues control (the correction factor) vector which is given by: $u_n = -k_p sgn(S)$, Where k_p is a controlled gain, sgn is the function $sign(S) = \frac{S}{|S|}$

The sliding mode surface S form takes the following form:

$$S = \sigma = (\frac{d}{dt} + k)^{r-1} e \qquad (2.4)$$

Where r is the degree relative of the system, e is the errors between the output vector and the desired reference input; k is a positive parameter which can be chosen arbitrarily or using a simple method that could lead to the appropriate choice.

2.1.1 SISO Sliding Mode Control

The sliding mode control law for SISO systems can be obtained using the equivalent and switching control law as follows:
The equivalent control law: is given by setting the derivative of the sliding mode surface to zero as follows:

$$\dot{S} = 0 \qquad (2.5)$$

And then solve for the control input law u which will be the equivalent control
The switching Control is given by following equation:

$$u_n = \dot{S} = -k_p sgn(S) \qquad (2.6)$$

Therefore, the overall control law $u = u_{eq} + u_n$ is then given by setting:

$$\frac{d}{dt}((\frac{d}{dt} + k)^{r-1} e) = -k_p sgn(S) \qquad (2.7)$$

And solve for the control law u, the idea is better illustrated using examples

2.1.2 MIMO Sliding Mode Control

Consider the MIMO nonlinear system given by:

$$\dot{x} = f(x) + \sum_{i=1}^{n} g_i u_i$$

$$y_1 = h_1(x)$$
$$y_2 = h_2(x) \tag{2.8}$$

$$\vdots$$

$$y_n = h_n(x) \tag{2.9}$$

The nonlinear system (2.8) has a vector relative degree $r_1, \ldots, r_n$ and has a vector output h_i. We assign a sliding mode surface for each output; a sliding mode control for MIMO systems are obtained by following the steps outlined for SISO systems.

The MIMO sliding mode control law is obtained by setting the derivatives of the sliding mode surfaces equal to the switching surfaces as shown:

$$\dot{S}_1 = -K_{p_1} sgn(S_1)$$
$$\dot{S}_2 = -K_{p_2} sgn(S_2)$$

$$\vdots \tag{2.10}$$

$$\dot{S}_n = -K_{p_n} sgn(S_n)$$

substituting the sliding mode surface given in (2.4) onto (2.10) gives:

$$\frac{d}{dt}(\frac{d}{dt} + k_1)^{r_1-1} e_1) = -K_{p_1} sgn(S_1)$$
$$\frac{d}{dt}(\frac{d}{dt} + k_2)^{r_2-1} e_2) = -K_{p_2} sgn(S_2)$$

$$\vdots \tag{2.11}$$

$$\frac{d}{dt}(\frac{d}{dt} + k_n)^{r_n-1} e_n) = -K_{p_n} sgn(S_n)$$

Where $k_1, k_2, \ldots k_n$ and $k_1, k_{p2}, \ldots k_{pn}$ are a controlled parameters, $e_1, e_2, \ldots e_n$ are the output errors of the system. The control law is then obtained by solving Equation 2.11 for the vector control u_i.

2.2 SLIDING MODE CONTROL EXAMPLES

2.2.1 Van der pol system

Consider the Van der Pol dynamical equations [4]:

$$\dot{x}_1 = x_2$$
$$\dot{x}_2 = -x_1 + \epsilon(1 - x_1^2)x_2 + u \qquad (2.12)$$
$$y = x_1$$

Find the sliding mode controller for the Van der Pol system.

2.2.1.1 Solution Using Hand

The sliding mode surface is then defined by:

$$S = (\frac{d}{dt} + k)^{r-1}e \qquad (2.13)$$

Where the error $e = (x_{1r} - x_1)$ is written in terms of the input reference x_{1r} and the output state x_1; k is a controlled parameter; the relative degree $r = 2$ in our problem. Therefore, the sliding mode surface of Equation 2.13 will be:

$$S = (\frac{d}{dt} + k)^{2-1}e$$
$$= \dot{e} + ke \qquad (2.14)$$
$$= (\dot{h}_{1r} - \dot{x}_1) + k(h_{1r} - x_1)$$

To simpliy the computation, we use the lie derivative notion as follows:

$$\dot{e} = \dot{h}_{1ref} - \dot{h}_1 \qquad (2.15)$$
$$= \dot{h}_{1ref} - \frac{\partial h_1}{\partial x}\dot{x}$$
$$= \dot{h}_{1ref} - \frac{\partial h_1}{\partial x}f(x) \qquad (2.16)$$

Where

$$L_f h(x) = \frac{\partial h_1}{\partial x}f(x) \qquad (2.17)$$

In order to find the sliding mode control law u, Equation 2.7 will be used:

$$\frac{d}{dt}(\dot{e} + ke) = -k_p sgn(S)$$

$$\ddot{e} + k\dot{e} = -k_p sgn(S) \quad (2.18)$$

$$(\ddot{h}_{1r} - L_f^2 h_1 - L_g L_f h_1 u) + k(\dot{h}_{1r} - L_f h_1) = -k_p sgn(S)$$

Where

$$L_f h_1 = x_2$$
$$L_f^2 h_1 = -x_1 + \epsilon(1 - x_1^2)x_2 \quad (2.19)$$
$$L_g L_f h_1 = 1$$

Substitute Equations 2.19 onto equation 2.18:

$$(\ddot{h}_{1r} - \dot{x}_2) + k(\dot{x}_{1r} - x_2) = -k_p sgn(S)$$
$$(\ddot{h}_{1r} - (-x_1 + \epsilon(1 - x_1^2)x_2 + u)) + k(\dot{h}_{1r} - \dot{x}_1) = -k_p sgn(S)$$
$$(2.20)$$

Thus, the sliding mode control law u is obtained from Equation 2.20 by:

$$u = \ddot{h}_{1r} + x_1 - \epsilon(1 - x_1^2)x_2 + k(\dot{h}_{1r} - \dot{x}_1) + k_p sgn(S) \quad (2.21)$$

2.2.1.2 Solution Using MATLAB®

```
The Nonlinear systems should be written in the following
form—Sliding Mode Controller for a class of Nonlinear
        systems—State space equations
        x=f(x)+gu
```

```
Number of states:=2
Number of inputs:=1
Parameters  e

parameters =

e

The vector f(x):=[x2;−x1+e*(1−x1^2)*x2]
The vector g(x):=[0;1]
The output variables:=x1
The sliding mode control law for SISO systems=:
```

```
Uc =

d3yr + x1 + kp*sgnS + k1*(d2yr - x2) + e*x2*(x1^2 - 1)
```

The Sliding mode surface:=

```
Surf =

d2yr - x2 + k1*(d1yr - x1)
```

The derivative of Sliding mode surface:=

```
dSurf =

d3yr + x1 + k1*(d2yr - x2) + e*x2*(x1^2 - 1)
```

2.2.2   DC motor angular position control

The mathematical model of the DC motor [16] is given by the following state space representation:

$$\dot{x}_1 = x_2 \tag{2.22}$$

$$\dot{x}_2 = -\frac{b}{J}x_2 + \frac{k_m}{J}x_3 \tag{2.23}$$

$$\dot{x}_3 = -\frac{k_e}{L}x_2 - \frac{R}{L}x_3 + \frac{1}{L}u$$

$$h_1 = x_1 \tag{2.24}$$

Where x_1 and x_2 are the angular position and velocity of the shaft respectively, x_3 is the current of the armature coil. u is the input voltage of the armature coil. The parameters are the resistance and inductance of the armature coil respectively, k_e and k_m are the speed and torque constants, b is the viscous friction and J is the moment of inertia. We seek to control the angular position using sliding mode control.

2.2.2.1   Solution Using Hand

The sliding mode surface is defined as:

$$S = (\frac{d}{dt} + k)^{r-1}e \tag{2.25}$$

Where the degree relative $r = 3$ in this example; the error $e = h_{1ref} - h_1$, Equation 2.25 will be:

$$S = \ddot{e} + 2k\dot{e} + e \tag{2.26}$$

The control law is given by differentiating the surface S:

$$\dot{S} = \dddot{e} + 2k\ddot{e} + k^2\dot{e} \tag{2.27}$$

Using the notion of lie derivative the errors in Equation 2.27 will be:

$$
\begin{aligned}
e &= h_{1ref} - h_1 \\
\dot{e} &= \dot{h}_{1ref} - \dot{h}_1 = \dot{h}_{1ref} - L_f h(x) \\
\ddot{e} &= \ddot{h}_{1ref} - \ddot{h}_1 = \ddot{h}_{1ref} - L_f^2 h(x) \\
\dddot{e} &= \dddot{h}_{1ref} - \dddot{h}_1 = \dddot{h}_{1ref} - L_f^3 h(x) - L_g L_f^2 h(x) u
\end{aligned} \tag{2.28}
$$

After having defined the derivative of the sliding mode surface, then Equation 2.7 will be used to construct the sliding mode surface:

$$\dddot{e} + 2k\ddot{e} + k^2\dot{e} = -k_p sgn(S) \tag{2.29}$$

Substituting Equation 2.28 onto 2.29 yields:

$$
\begin{aligned}
&\dddot{h}_{1ref} - L_f^3 h(x) - L_g L_f^2 h(x) u + 2k(\ddot{h}_{1ref} - L_f^2 h(x)) \\
&+ k^2(\dot{h}_{1ref} - L_f h(x)) = -k_p sgn(S)
\end{aligned} \tag{2.30}
$$

The control law u is then given by:

$$
\begin{aligned}
L_g L_f^2 h(x) u = &\dddot{h}_{1ref} - L_f^3 h(x) + 2k(\ddot{h}_{1ref} - L_f^2 h(x)) \\
&+ k^2(\dot{h}_{1ref} - L_f h(x)) + k_p sgn(S)
\end{aligned} \tag{2.31}
$$

Where

$$
\begin{aligned}
L_f h(x) &= x_2 \\
L_f^2 h(x) &= -\frac{b}{J} x_2 + \frac{k_m}{J} x_3 \\
L_f^3 h(x) &= \frac{1}{J} \left(b \frac{(bx_2 - k_m x_3)}{J} \right) + \frac{1}{J} \left(k_m \frac{(Rx_3 + k_e x_2)}{L} \right) \\
L_g L_f^2 h(x) &= \frac{1}{L}
\end{aligned} \tag{2.32}
$$

Substituting Equation 2.32 onto Equation 2.31, the sliding mode control law is:

$$
\begin{aligned}
u = &\dddot{h}_{1ref} + \frac{k_e}{L} x_2 + \frac{R}{L} x_3 \\
&+ 2k(\ddot{h}_{1ref} - \frac{1}{J}(b\frac{(bx_2 - k_m x_3)}{J}) + \frac{1}{J}(k_m\frac{(Rx_3 + k_e x_2)}{L})) \\
&+ k^2(\dot{h}_{1ref} - x_2) + k_p sgn(S)
\end{aligned} \tag{2.33}
$$

2.2.2.2 Solution Using MATLAB

The Nonlinear systems should be written in the following
form—Sliding Mode Controller for a class of Nonlinear
           systems—State space equations
           x=f(x)+gu

Number of states:=3
Number of inputs:=1
Parameters b J km ke L R

parameters =

b J km ke L R

The vector f(x):=[x2;-(b/J)*x2+(km/J)*x3;
     -(ke/L)*x2-(R/L)*x3]
The vector g(x):=[0;0;1/L]
The output variables:=x1
The sliding mode control law for SISO systems=:

Uc =

d4yr + kp*sgnS + k1*(d3yr + (b*x2)/J - (km*x3)/J)
    + k2*(d2yr - x2) + (km*((R*x3)/L + (ke*x2)/L))/J
    - (b*((b*x2)/J - (km*x3)/J))/J

The Sliding mode surface:=

Surf =

d3yr + k2*(d1yr - x1) + k1*(d2yr - x2) + (b*x2)/J
    - (km*x3)/J

The derivative of Sliding mode surface:=

dSurf =

d4yr + k1*(d3yr + (b*x2)/J - (km*x3)/J) + k2*(d2yr - x2)
    + (km*((R*x3)/L + (ke*x2)/L))/J-(b*((b*x2)/J
    - (km*x3)/J))/J

2.2.3 Permanent Magnet Synchronous motor speed control

The dynamical model of PMSM [15] describing the electrical and mechanical part is given by:

$$
\begin{aligned}
\dot{x}_1 &= -\frac{R}{L_d}x_1 + p\frac{L_q}{L_d}x_2x_3 + \frac{1}{L_d}u_d \\
\dot{x}_2 &= -\frac{R}{L_q}x_2 - p\frac{L_q}{L_d}x_1x_3 - p\frac{\Phi}{L_q} + \frac{1}{L_q}u_q \\
\dot{x}_3 &= p\frac{\Phi_f}{J}x_2 - p\frac{L_q - L_d}{J}x_1x_2 - \frac{f}{J}x_3 - \frac{1}{J}\tau
\end{aligned}
\tag{2.34}
$$

Where x_1 and x_2 are the d and q axis stator currents respectively; x_3 is the mechanical speed of the motor; u_d and u_q are the d axis and q axis stator voltages respectively; R and $L_d = L_d$ are the winding resistance and inductance on the d and q axis. J is mechanical inertia of the motor; τ is the electrical torque.

The objective is to control the mechanical velocity x_3 and the x_1 current.

$$
h(x) = \begin{bmatrix} h_1 \\ h_2 \end{bmatrix} = \begin{bmatrix} x_1 \\ x_3 \end{bmatrix}
\tag{2.35}
$$

2.2.3.1 Solution Using Hand

We note that the relative degrees of the current and the mechanical velocity are $r_1 = 1$ and $r_2 = 2$, respectively. Therefore, the sliding mode surfaces are:

$$
S_1 = (\frac{d}{dt} + k)^{r_1-1}e_1 = e_1
\tag{2.36}
$$

$$
S_2 = (\frac{d}{dt} + k)^{r_2-1}e_2 = \dot{e}_2 + ke_2
$$

Where the errors are $e_1 = h_{1ref} - h_1$ and $e_2 = h_{2ref} - h_2$. Thus, the control law is then obtained by differentiating the sliding mode surfaces in Equation 2.37 and equating to (2.6), yields:

$$
\dot{e}_1 = -k_{p1}sgn(S_1)
\tag{2.37}
$$

$$
\ddot{e}_2 + k\dot{e}_2 = -k_{p2}sgn(S_2)
$$

Using the notion of lie derivative Equation 2.37 will be:

$$
\dot{h}_{1ref} - L_f h_1 - L_{g_1} h_1 u_d = -k_{p1}sgn(S_1)
\tag{2.38}
$$

$$
\ddot{h}_{2ref} - L_f^2 h_2 - L_{g_1} L_f h_2 u_d - L_{g_2} L_f h_2 u_q + k(\dot{h}_{2ref} - L_f h_2)
$$

$$
= -k_{p2}sgn(S_2)
$$

Rearranging Equation 2.38 and solve for the control law u_d and u_q gives:

$$\begin{bmatrix} u_d \\ u_q \end{bmatrix} = \begin{bmatrix} L_{g1}L_f h_1 & 0 \\ L_{g_1}L_f h_2 & L_{g_2}L_f h_2 \end{bmatrix}^{-1}$$
$$\begin{bmatrix} \dot{h}_{1ref} - L_f h_1 + k_{p1}sgn(S_1) \\ \ddot{h}_{2ref} - L_f^2 h_2 + k(\dot{h}_{2ref} - L_f h_2) + k_{p2}sgn(S_2) \end{bmatrix} \quad (2.39)$$

Where:

$$\begin{bmatrix} L_{g1}L_f h_1 & 0 \\ L_{g_1}L_f h_2 & L_{g_2}L_f h_2 \end{bmatrix} = \begin{bmatrix} \frac{1}{L_d} & 0 \\ \frac{px_2(L_d-L_)}{JL_d} & \frac{p\Phi+px_1(L_d-L_q)}{JL_d} \end{bmatrix} \quad (2.40)$$

$$L_f h_1 = -\frac{R}{L_d}x_1 + p\frac{L_q}{L_d}x_2 x_3$$
$$L_f h_2 = p\frac{\Phi_f}{J}x_2 - p\frac{L_q - L_d}{J}x_1 x_2 - \frac{f}{J}x_3 - \frac{1}{J}\tau$$
$$L_f^2 h_2 = (f\frac{\tau + fx_3 - p\Phi x_2 - *x_1 x_2(L_d - L_q)}{J^2}) \quad (2.41)$$
$$= -\frac{(p\Phi + px_1(L_d - L_q))(Rx_2 + p\Phi + pL_q x_1 x_2)}{JL_d}$$
$$= -\frac{px_2(L_d - L_q)(Rx_1 - pL_q x_2 x_3)}{JL_d}$$

2.2.3.2 Solution Using MATLAB

The Nonlinear systems should be written in the following
form——Sliding Mode Controller for MIMO
            Nonlinear systems——State space equations
            x=f(x)+gu

```
Number of states:=3
Number of inputs:=2
Parameters  R Ld Lq phi f J p Tl
The vector f(x):=[-(R/Ld)*x1+p*(Ld/Lq)*x2*x3;
                 -(R/Lq)*x2-p*(Lq/Ld)*x1*x3-p*phi/Ld;
                 p*(phi/J)*x2-p*((Lq-Ld)/J)*x1*x2
                 -(f/J)*x3-Tl/J]
The vector g(x):=[(1/Ld) 0;0 (1/Lq);0 0]
The output variables:=[x1;x3]
The relative degree of  h1
equal:=1
```

```
The relative degree of  h2
equal:=2
The sliding mode control law for MIMO systems
————Is Given by Uc=:inv(Lhg)*(S)————
————The function S:=————

S= d2yr1 + kp1*sgnS1 + (R*x1)/Ld — (Ld*p*x2*x3)/Lq ;
   d2yr2 + kp2*sgnS2 + Tl/J + k1*(d3yr2 + ((p*phi)/J
   + (p*x1*(Ld — Lq))/J)*((R*x2)/Lq + (p*phi)/Ld
   + (Lq*p*x1*x3)/Ld) — (f*(Tl/J + (f*x3)/J — (p*phi*x2)/J
   — (p*x1*x2*(Ld — Lq))/J))/J + (p*x2*(Ld — Lq)*((R*x1)/Ld
   — (Ld*p*x2*x3)/Lq))/J) + (f*x3)/J — (p*phi*x2)/J
   — (p*x1*x2*(Ld — Lq))/J


————The Matrix Lhg:=————

Lhg =

[              1/Ld,                              0]
[ (p*x2*(Ld — Lq))/(J*Ld), ((p*phi)/J
  + (p*x1*(Ld — Lq))/J)/Lq]

The sliding mode control law for MIMO systems=:

Uc =Ld*(d2yr1 + kp*sgnS +(R*x1)/Ld — (Ld*p*x2*x3)/Lq);

(J*Lq*(d2yr2 + kp*sgnS + Tl/J + k1*(d3yr2 + ((p*phi)/J
+ (p*x1*(Ld — Lq))/J)*((R*x2)/Lq + (p*phi)/Ld
+ (Lq*p*x1*x3)/Ld) — (f*(Tl/J + (f*x3)/J — (p*phi*x2)/J
— (p*x1*x2*(Ld — Lq))/J))/J + (p*x2*(Ld — Lq)*((R*x1)/Ld
— (Ld*p*x2*x3)/Lq))/J) + (f*x3)/J — (p*phi*x2)/J
— (p*x1*x2*(Ld — Lq))/J))/(p*(phi + Ld*x1 — Lq*x1))
— (Lq*x2*(Ld — Lq)*(d2yr1 + kp*sgnS + (R*x1)/Ld
— (Ld*p*x2*x3)/Lq))/(phi + Ld*x1 — Lq*x1)
```

2.3 TUNING OF SLIDING MODE CONTROL PARAMETER

The parameters of the sliding mode control law should be chosen in order to enhance the performance of the controlled system. However, those parameters are said to be chosen arbitrarily using trial-and-error method, this method may take a longer time to find approximately the exact values.

We will use a simple method that leads to approximately the exact values by identification with a characteristic of first or second order systems with the Equation 2.7. Let's suppose that we have a nonlinear system with a relative degree $r = 2$; therefore, Equation 2.7 will be:

$$\ddot{e} + k\dot{e} = -k_p sgn(S)$$

since the $sgn(s) = \frac{S}{|S|}$, equation will be:

$$\ddot{e} + k\dot{e} = -k_p \frac{S}{|S|}$$

$$\ddot{e} + k\dot{e} = -k_p \frac{\dot{e} + ke}{|S|} \tag{2.42}$$

We note that the $|S|$ is a positive function written in terms of the error, which can be chosen as a small value $|S| << 1$. Thus, Equation 2.42 will be:

$$\ddot{e} + (k + \frac{k_p}{|S|})\dot{e} + \frac{k}{|S|}e = 0 \tag{2.43}$$

Equation 2.43 takes the form of a second order system:

$$s^2 + 2\zeta\omega s + \omega^2 = 0 \tag{2.44}$$

By identification to Equation 2.44 to 2.43, yields:

$$(k + \frac{k_p}{|S|}) = 2\zeta\omega$$

$$\frac{k}{|S|} = \omega^2 \tag{2.45}$$

The variable ω and ζ are chosen to meet specific requirements

2.4 MATLAB-BASED FUNCTION: SLIDING MODE CONTROL

MATLAB programs listed on this chapter are used to find the sliding mode control law for SISO and MIMO systems using Symbolic MATLAB Library. To simplify the use for these programs a test function is developed. After calling this function, the user has to enter the necessary functions which are: $f(x)$, (g) and the output vector $h(x)$ using symbolic variables. The principle MATLAB functions are:

NonContSidFed: This function calculates the lie derivatives of the given SISO dynamical nonlinear system

SlidingModeTerms: This function calculates the sliding mode terms like the surface and its derivative, as well as the control law U

The derivatives of the output y for SISO nonlinear systems are described by the following variables:

$$\left[\begin{array}{cccc} y & y' & .... & y^n \end{array}\right] = \left[\begin{array}{cccc} d1yr & d2yr & .... & d(n+1)yr \end{array}\right] \quad (2.46)$$

MIMOSlidingModeLieDer: This function calculates the lie derivatives for MIMO dynamical nonlinear systems

MIMOSlidingModeController: This function calculates the sliding mode terms and control law for MIMO nonlinear systems

The derivatives of the output y for MIMO nonlinear systems are described by the following variables:

$$\begin{bmatrix} y_1 & y_1' & .... & y_1^n \\ y_2 & y_2' & .... & y_2^n \\ \vdots & & & \\ y_n & y_n' & .... & y_n^n \end{bmatrix} = \begin{bmatrix} d1yr1 & d2yr1 & .... & d(n+1)yr1 \\ d1yr2 & d2yr2 & .... & d(n+1)yr2 \\ \vdots & & & \\ d1yrn & d2yrn & .... & d(n+1)yrn \end{bmatrix}$$

$$(2.47)$$

2.5 MATLAB PROGRAMMES DESCRIPTION

The developed programmes are used to find the sliding mode control law for any types of nonlinear systems (SISO and MIMO) using MATLAB symbolic library. Hence, the user will put three separate MATLAB codes on the same folder which are:

MATLAB name functions of SISO systems

- Testfunction
- NonContSidFed
- SlidingModeTerms

MATLAB name functions of MIMO systems

- Testfunction
- MIMOSlidingModeLieDer
- MIMOSlidingModeController

After calling the test functions by users, they will be asked to introduce the system functions $f(x)$, $g(x)$ and the output vector $h(x)$ as symbolic functions. Therefore, each program from those listed above will output useful functions which are:

(1) The functions MIMOSlidingModeLieDer and NonContSidFed will give us for variables which are:

    (a) The lie derivatives of h along the vector field f and g which are represented by: Lhf and the decoupling matrix Lhg

    (b) The vector of all lie derivatives along the vector field g

    (c) The relative degree r of the systems

(2) The functions MIMOSlidingModeController and SlidingModeTerms will give us for variables which are:

    (a) The MIMOSlidingModeController will provide as with: $[e, der, Surf, dSurf, Uc, S]$, where e is the errors vector, der: represent the derivatives vector, Surf: is the sliding mode surface; dSurf: represent the derivative of the surface; S is a vector contain the switching terms such that the controller Uc is equal to $Uc = inv(Lhg) * S$

    (b) $[Surf, dSurf, dd, K, Uc] = SlidingModeTerms(Hc, L, r, Lhg)$; Surf is the surface of a SISO system; dSurf: The derivative of the surface; K: The controller parameter vector; Uc: the control law for SISO systems

2.6 MATLAB CODES FOR SISO SLIDING MODE

2.6.1 MATLAB test function for SISO systems

```
clear all;clc
disp('———————————————————————————');
disp('The Nonlinear systems should be written in the
        following form');
disp('——Sliding Mode Controller for a class of Nonlinear
        systems——');
disp('    State space equations    x=f(x)+gu        ');
disp('———————————————————————————');
n=input('Number of states:=');
nin=input('Number of inputs:=');
x=sym(zeros(1,n));
u=sym(zeros(1,nin));

for j=1:n
    eval(sprintf('syms x%d',j))
    x(:,j)=sprintf('x%d',j);
end
for k=1:nin
    eval(sprintf('syms u%d',k));
```

```
    u(:,k)=sprintf('u%d',k);
end
syms u1
f=input('The vector f(x):=','s');
g=input('The vector g(x):=','s');
Hc=input('The output variables:=','s');
[Lhf,Lhg,dh,L,u,r]=NonContSidFed(f,g,Hc,x,u1);
[Surf,dSurf,dd,K,Uc]=SlidingModeTerms(Hc,L,r,Lhg);
```

2.6.2 MATLAB function for SISO systems

```
function [Lhf,Lhg,dh,L,u,r]=NonContSidFed(f,g,Hc,x,v)
% The NonContSidFed MATLAB function finds the lie
% derivatives and the decoupling matrix that has to be used
% to find the Sliding mode controller, by providing this
% function with:
% f        : The system function f(x)
% g        : The system output function g(x)
% x        : The state vector x=[x1,x2,....,xn]
% v        : The new input vector
% Hc       : The output of the system hn(x)=xn
% The program has to give the user the following functions
% Lhf      : The lie derivative of the vector h(x) along
%              f(x) Lhf=[L^{r-1}hfn]
% Lhg      : The lie derivative of the vector field h
%              along the function g; Lhg=[Lg1Lfhn]
% L        : The lie derivatives L=[Lhf1 L^{2}hf1.......
% L^{r-1}hf1]
% x=f(x)+gu
L=[];
Lhf=[];Lhg=[];
if(nargin<4)
    error('Not enough input arguments!');
end
n1=length(Hc);
n=length(x);
k=1;kk=1;
while(k<n+1)
    dh=LieDer(Hc,x,n);
    Lg=dh*g;
    L=[L,dh*f];
    if(Lg~=0)
        r=k;
        Lhf=[Lhf;dh*f];
        Lhg=[Lhg;Lg];
      · break;
    end
    Hc=dh*f;
    k=k+1;
end
u=inv(Lhg)*(Lhf-v);
function dh=LieDer(Hc,x,n)
```

```
dh=[];
for ii=1:n
    d=diff(Hc,x(ii));
    dh=[dh,d];
end
end
r=k;
end


function [Surf,dSurf,dd,K,Uc]=SlidingModeTerms(Hc,L,r,Lhg)
% The SlidingModeTerms MATLAB function finds the sliding mode
% surface and its derivatives for SISO nonlinear systems
% That takes the following form:
%       xp=f(x)+gu
%        y=h(x)
% Hc       : The output of the system Hc=x
% L        : The lie derivatives L=[Lhf1 L^2hf1.......
%             L^{r-1}hf1]
% r        : The relative degree of the system r=r1
% Surf     : Sliding Mode surface for SISO systems
%             Surf=(d/dt+k)^(r-1)e
% dSurf    : The derivative of the sliding surface
%             Surf=d((d/dt+k)^(r-1)e)/dt
% K        : The controller parameter
%             vector K=[1 k1 k2,....kn]
k=[];
syms kp Uc U sgnS
if (r==1)
    K=1;
else
    k=sym(zeros(1,r-1));
    for jj=1:r-1
    eval(sprintf('syms k%d',jj));
    k(:,jj)=sprintf('k%d',jj);
    end
    K=[1 k];
end

dd=sym(zeros(1,r));
for ii=1:r+1
    eval(sprintf('syms d%dyr',ii));
    dd(:,ii)=sprintf('d%dyr',ii);
end
e=[dd(1)-Hc];
%s=[dd(r)-Hc];
s=[];
k=1;
for kk=2:r
    e=[e;(dd(kk)-L(k))];
    k=k+1;
end
LL=fliplr(L);
```

```
dl=fliplr(dd);

for kk=1:1:r
    s=[s;(dl(kk)-LL(kk))];
end
S=flipud(e);
Sp=s;
Uc=Lhg*U;
disp(['The sliding mode control law for SISO systems=:']);
Uc=K*Sp+kp*sgnS
disp(['The Sliding mode surface:=']);
Surf=K*S
disp(['The derivative of Sliding mode surface:=']);
dSurf=K*Sp
```

2.7 MATLAB CODES FOR MIMO SLIDING MODE

2.7.1 MATLAB test functions for MIMO systems

```
clear all;clc
disp('─────────────────────────────────────');
disp('The Nonlinear systems should be written in the
        following form');
disp('──Sliding Mode Controller for MIMO Nonlinear
        systems──');
disp('              State space equations    x=f(x)+gu
');
disp('─────────────────────────────────────');
n=input('Number of states:=');
nin=input('Number of inputs:=');
x=sym(zeros(1,n));
u=sym(zeros(1,nin));

par=input('Parameters   ','s');
eval(sprintf('syms %s',par));
parameters=sprintf('%s',par);

for j=1:n
   eval(sprintf('syms x%d',j))
   x(:,j)=sprintf('x%d',j);
end
for k=1:nin
    eval(sprintf('syms u%d',k));
    u(:,k)=sprintf('u%d',k);
end

f=input('The vector f(x):=','s');
g=input('The vector g(x):=');
h=input('The output variables:=','s');
f=sym(f);h=sym(h);g=sym(g);
[Lhf Lhg L r]=MIMOSlidingModeLieDer(f,g,h,x);
[e,der,Surf,dSurf,Uc]=MIMOSlidingModeController(h,L,r,Lhg);
```

2.7.2 MATLAB functions for MIMO systems

```
function [Lhf Lhg L r]=MIMOSlidingModeLieDer(fx,g,h,x)
% This programme finds a nonlinear controller which is the
% input output feedback linearization controller where the
% system should be given in symbolic format this  function
% works well with SISO and MIMO systems which is written in
% compact form like       dx=f(x)+gu;
%                  y=h(x);
% x                : The state vector x=[x1,x2,...,xn]
% h                : The output vector h(x)=[x1;x2;...;xn]
% fx               : The f(x) the describes the system
% g                : The g.u vector of the output system
% u                : The output vector
% Lhf              : The lie derivative of the vector file
%          .         h along the function f(x)
%                    Lhf=[L^{r1-1}fh1;L^{r2-1}fh2;...;
%                    L^{rn-1}fhn]
% Lhg              : The lie derivative of the vector
%                    field h along the function g
%                    Lhg=[Lg1Lfh1, Lg2Lfh1,.....,LgnLfh1;
%                        .........;
%                        Lg1Lfhn,Lg2Lfhn,......, LgnLfhn]
% r                : The degree relative of the system
%                    r=[r1;r2;........;rn]
% L                : The lie derivative vector
%                  : L=[Lhf1 Lhf2....... Lhfn;
%                      L^2hf1 L^2hf2....L^2hfn;
%
%                      L^(r1-1)hf1 L^(r2-1)hf2.....
%                  L^(rn-1)hfn]
% ]
if nargin <4
    error('Not enough input argument');
end
k=1;L=[];kk1=0;
Lhg=[];Lhf=[];vc=1;
nb=length(h);
r=zeros(1,nb);
while k<length(h)+1
h1=h(k);
for i=1:nb+1
df=Lie_Derivative(h1,x);           % this Lie derivative
                                   % function
[lhf lhg]=solve_lie_der(df,fx,g); % solve for the g
L=[L,lhf];
[n b]=size(lhg);
for ii=1:n
  d=any(lhg(ii,:) ~= 0);
end
if d==1;
    disp(['The relative degree of  h',num2str(k)]);
```

```
        disp(['equal:=',num2str(i)]);
        r(1,k)=i;
        break;
    else
        h1=lhf;

    end
    if i==nb+1 && d==0
        disp(['The system dose not admit an input output
        feedback linearization']);
        return;
    end

end
Lhg=[Lhg;lhg];
Lhf=[Lhf;lhf];
k=k+1;
end
function df=Lie_Derivative(h,x)
% The LieDerivative MATLAB function is used
% to find the jacobian vector of a given output
% h        : Is the output function h(x)=[x1;x2;...;xn]
% x        : The state vector x=[x1,x2,.....,xn]
% df       : The jacobian of h along x
if nargin<2 & nargin==0
    error('not enough input argument');
end
df=[];
n=length(x);
for ii=1:n
    xx=x(ii);
    dff=diff(h,xx);
    df=[df,dff];
end
df;
end
function [lhf lhg]=solve_lie_der(df,fx,G)
% This equation solves the Lie derivativies that is
% multiplied by
% The output vectors Lhg=Lh*g*u where
% LH       : The Lie derivativies of h(x) along the vector
%            field fx
% fx       : The vector field f(x) that describes the system
% G        : The vector field of the input
LHg=[];
lhf=df*fx;
[n,b]=size(G);
for ii=1:b
    Lgh=df*G(:,ii);
    LHg=[LHg,Lgh];
end
lhg=LHg;
end
```

```
end


function [e,der,Surf,dSurf,Uc,S]=MIMOSlidingModeController
    (h,L,r,Lhg)
% The mimo nonlinear systems described by the following
%                     xp=f(x)+gu
% e       : Is the error vector
%             e=[e1 e2 ....en;de1 de2 ....den;.....;dne1
%             dNe.... dNen]
% der     : The derivative of the output vector
%             der=[h1 h2 ...hn;d1h1 d1h2....dNhn;......;dNh1
%             dNe....dNen]
% Surf    : The Sliding Mode Surface for MIMO systems
%             Surf=(d/dt+k)^(r-1)e
% dSurf   : The derivative of the sliding mode surface
%             dSurf=d((d/dt+k)^(r-1)e)/dt
% Uc      : The Sliding mode control law
% h       : The output vector y=[h1;h2;....;hN] for MIMO
%             systems
% L       : The lie derivative vector of the MIMO system
%             L=[Lhf1 Lhf2....... Lhfn;
%                 L^2hf1 L^2hf2....L^2hfn;
%
%                   L^rhf1 L^rhf2..... L^rhfn]
% r       : The relative degree vector r=[r1,r2,r3,....,rn]
% Lhg     : The lie derivative of h along the vector field g
%             Lhg=[Lg1Lfh1, Lg2Lfh1,.....,LgnLfh1;
%                   .........;
%                   Lg1Lfhn,Lg2Lfhn,......, LgnLfhn]
e=[];kk=1;
rr=max(r);
der=[];S=[];Surf=[];Ss=[];
nb=length(h);d=[];qq=1;q=1;
if (r==1)& (nb==1)
    error('The output vector h should be of length >=2');
end

if nargin <4
    error('Not enough input argument');
end
LL=sym(zeros(rr+1,nb));
der=sym(zeros(rr+1,nb));


syms Uc U
LL(1,:)=h;

k=[];
K=sym(zeros(nb,rr));
KK=sym(zeros(nb,rr+1));
KK(:,1)=sym('1');
%K(:,1)=1;
```

```matlab
for ll=1:nb
    R=r(ll);
    k=sym(zeros(1,R-1));
    for jj=1:R-1
    eval(sprintf('syms k%d',jj));
    k(:,jj)=sprintf('k%d',jj);
    end
    K(ll,:)=[1 k];
end
KK(:,nb:rr+1)=K;
F=subs(KK(:,nb+1:rr+1),{1},{0});
KK(:,nb+1:rr+1)=F;
%% The functions sgnS1, Sgn(S2),....sgn(Sn)
    % The parameters kp1, kp2,....kp3
 sgns=sym(zeros(1,nb));
 kp=sym(zeros(1,nb));
 for jj=1:nb
    eval(sprintf('syms sgnS%d',jj));
    sgns(:,jj)=sprintf('sgnS%d',jj);
 end
    for jj=1:nb
    eval(sprintf('syms kp%d',jj));
    kp(:,jj)=sprintf('kp%d',jj);
    end
%%
for jj=1:nb
    R=r(jj);
dd=sym(zeros(1,R));
for ii=1:R+1
    eval(sprintf('syms d%dyr%d',ii,kk));
    dd(:,ii)=sprintf('d%dyr%d',ii,kk);
end
d=[d,dd];

kk=kk+1;

end

for jj=1:nb
    R=r(jj);
for mm=2:R+1
    LL(mm,jj)=L(qq);
    qq=qq+1;
end
for nn=1:R+1
    der(nn,jj)=d(q);
    q=q+1;
end
end
```

```
e=flipud(der—LL);

for bb=1:nb
    Kd=fliplr(KK(bb,nb:rr+1));
    ee=e(1:rr,bb);
    nm=kp(bb)*sgns(bb);
    sc=Kd*ee+kp(bb)*sgns(bb);
    S=[S;sc];
    Ss=[Ss;Kd*ee];
end

es=flipud(e);
for cc=1:nb
    R=r(cc);
    dK=fliplr(KK(cc,nb:R+1));
    esf=(es(1:R,cc));
    surface=dK*(esf);
    Surf=[Surf;surface];
end
dSurf=Ss;
Surf=Surf;
[nh bh]=size(Lhg);
disp(['The sliding mode control law for MIMO systems']);
disp(['————Is given by Uc=:inv(Lhg)*(S)——————————']);
disp(['————The function S:=——————']);
S
disp(['————The Matrix Lhg:=——————']);
Lhg

for ii=1:nh
  invg=any(Lhg(ii,:)~=0);
end
if(invg~=0)
disp(['The sliding mode control law for MIMO systems=:']);
Uc=inv(Lhg)*(S)
else
disp(['The sliding mode controller is not possible']);
disp(['for this output vector try to choose another one!']);
end
```

Bibliography

[1] Shankar S. (1999). Nonlinear systems analysis, stability and control. *Springer, Interdisciplinary applied mathematics*. New York, 1999.

[2] A. Isidori. (1989). Nonlinear control systems. Second Ed. Springer Verlag. Berlin, Heidelberg.

[3] Knight A. (2000). Basics of MATLAB and Beyond. Chapman Hall/CRC Press. Boca Raton, FL.

[4] Khalil, H. K. (1992). Nonlinear systems. Englewood Cliffs, NJ: Prentice Hall.

[5] C Edwards, S Spurgeon. (1998). Sliding mode control: Theory and applications. CRC Press

[6] Shihua Li, Xinghuo Yu, Leonid Fridman, Zhihong Man, Xiangyu Wang, Advances in Variable Structure Systems and Sliding Mode Control-Theory and Applications, Springer International Publishing 2018

[7] S. V. Emelyanov, Variable-Structure Control Systems [in Russian], Nauka, Moscow (1967).

[8] Yuri Shtessel, Christopher Edwards, Leonid Fridman. Sliding Mode Control and Observation. Birkhauser Boston Inc (2013)

[9] Minsun Kim.Tae-Yong Kuc. Hyosin Kim Seok-min Wi.Jin S Lee: An adaptive learning controller for MIMO uncertain feedback lin-earizable nonlinear systems: Nonlyniear Dyn (2015) Springer Sci-ence

[10] Lei-Po Liu Zhu-Mu Fu, Xiao-na Song Sliding mode con-trol with disturbance observer for a class of nonlinear systems. International journal of automation and computing,october 2012, 487

[11] Juan Fernandez Vargas, Gerard Ledwich, Variable structure control for power systems stabilization, International Journal of Electrical Power Energy System Vol 32, Issue 2, February (2010)

[12] Boufadene, M., Belkheiri, M. and Rabhi A.: Adaptive nonlinear observer augmented by radial basis neural network for a nonlinear sensorless control of an induction machine, Int. J. Automation and Control, Vol. 12, No. 1, pp.2743 (2018)

[13] Lei-Po, L., Zhu-Mu, F. and Xiao-na, S: Sliding mode control with disturbance observer for a class of nonlinear systems, International Journal of Automation and Computing, Vol. 9, No. 5, pp.487491 (2012)

[14] Andreas, M. Recent Advances in Robust Control Novel Approaches and Design Methods, InTech, Croatia (2011)

[15] htessel, Y., Edwards, C., Fridman, L., Levant, A.Sliding Mode Control and Observation (2014)

[16] Marwa EZZAT, Jesus DE LEON, Nicolas GONZALEZ and Alain GLUMINEAU. Sensorless Speed Control of Permanent Magnet Synchronous Motor by using Sliding Mode Observer, VSS (2010), Mexico City, Mexico, 26-28 June

[17] N.P.Quang, J. A.Dittrich: Vector Control of Three-Phase AC Machine Springer (2008)

[18] Q. Teng, J. Tian, J. Duan, H. Cui, J. Zhu and Y. Guo, "Sliding-mode MRA observer-based model predictive current control for PMSM drive system with DC-link voltage sensorless," 2017 20th International Conference on Electrical Machines and Systems (ICEMS), Sydney, NSW, 2017, pp. 1-6. doi: 10.1109/ICEMS.2017.8056033